The Truth of the Universe and Happiness

Kazushi Taira

Published by Kazushi Taira, 2023.

While every precaution has been taken in the preparation of this book, the publisher assumes no responsibility for errors or omissions, or for damages resulting from the use of the information contained herein.

THE TRUTH OF THE UNIVERSE AND HAPPINESS

First edition. September 28, 2023.

ISBN: 979-8215629017

Written by Kazushi Taira.

Table of Contents

(Introduction)

You can see the power of logic in Descartes' "I think, therefore I am," which proves the existence of the "Self" without the need for scientific measurement.

In this book, firstly in an article, I will try to prove the following truth: "The whole universe" (the whole world) is infinite, eternal, and transcends time with the power of logic.

I think this worldview could be the basis for universal happiness for everyone in "the whole universe," if each of us does not stick to our own standpoints. The word "whole" implies "holiness." I feel a mystery in the nature of the existence of "the whole universe."

I have included two rather easy-to-understand papers that originated the main thesis, which is rather difficult and rigid in style. The two papers are placed before the main thesis in the first part (I).

Secondly in an essay (II), I will try to explain what happiness is, and how one can be happy.

Then, some of my other essays and articles will be included in the third part (III).

I hope this book will be your light in the darkness.

Prologue to the Second Edition

In this second edition, I have revised some of the articles and rearranged the order of some of them. I have also added "Dialogue with the Finite Theory of the Universe, etc.".

I hope this will make my theory of the infinity of the universe easier to understand.

(I) The Truth of the Universe

In this part, I will try to explain that the whole universe in infinite, eternal and transcends time.

1-1: About Infinity

Is there an end of the universe?

There is no end. The expanse of the whole universe (everything) is infinite.

For example, if you travel the universe toward its end by spaceship, you cannot reach it, because the beyond inevitably exists. Even if the universe were contained within something like a wall, the wall is not the end. Because the layer of the wall must have its limit and the beyond inevitably exists.

Even if the expanse of the universe were like the inside of a balloon, the outside inevitably exists. Even if the universe were encircled within nothing, the expanse of the whole universe (everything) is infinite by including the nothing.

You might doubt this, but it becomes clear by logically thinking as below.

When you categorize the territory of the universe to any extent, the outsides inevitably exist. Therefore, "the whole expanse of the universe" (everything), including the close-to-absolute null (nothing) is infinite.

Let me explain this more precisely.

When you assume a certain category (let B stand for the category hereafter), you are assuming a category other than B at the same time. (Confer Set of mathematics), However, if someone doubt that the category does not exists, they may think that all categories do not exist, therefore everything is null!

Then, what can be proven if one category really exists? The answer is the existence of "everything" which consists of the existence of one category and the beyond.

Now I will prove the existence of one category using the words of Descartes.

His words "I think, therefore I am" tell us the following truth: If a subject does not exist, it cannot think; so the thinking ego really exists and is not a mere illusion. Therefore, the existence of self is proven, and the existence of the world where the self must exist at or within (the category B) is also proven.

In addition, to exist means to be a subject or object of acts or perception. If one does not exist, it cannot act nor receive anything. Thinking about playing catch leads to an understanding of it. To say it again, to exist means to be a subject or object of an act or perception.

Then, what is null? This question is important when we think about the existence of self and world.

The null is, precisely, an existence close to null. To put it simply, the null is one of the existence.

"The existence close to null" can be divided into two; "the relative null" and "the close-to-absolute null. "The relative null is an existence that can be an object to action or perception for some beings, while not being perceived for the others. For example, ultraviolet is the relative null that really exists, while it cannot be perceived by the five senses of human being. As a side note, ultraviolet can be perceived, by using a sensor, by human beings.

Meanwhile, the close-to-absolute null is a state which can only be perceived by logic.

Conversely speaking, it cannot be an object to action or perception by anything but logic. That is a "nothing" in common usage of words. But there is no phenomenon that cannot be objectified by logic. Sensuously explaining, "objectify the nothing", "give a name to "nothing"". ◈

In other words, we can give a name as "close-to-absolute null" to a state which cannot be objectified to act or perception except by logic. It exists as "nothing", not as a delusion nor illusion since is being perceived correctly. Therefore, a state in which any cannot be objectified or perceived does not exist: "the absolute null" is impossible. "The absolute null" is a misuse of logic.

I defined "existence" and "close to absolute null" as above.

Now back to the main question.

Is there an end of the universe?

There is no end. It is infinite.

when we categorize the universe at any point, there is always the beyond. Since the category where the self really exist is the universe, the existence of self and the universe is proven. At the same time, the existence of the beyond is also proven.

Therefore, the expanse of the whole universe (everything) is infinite.

Say it in other words, everything is a real and infinite world that consists of existence and existence close to absolute null.

Since the expanse is infinite, we cannot reach its end physically. However, we can exert, even if it might be little, the power of creation. The peace has not yet been fully realized even on the little earth in the vast universe. We must change the earth for the better by creative cooperation.

-

1-2: "Everything" and the Big Bang Theory

-

"Everything" will continue to exist forever, since it will not disappear. There is no beginning or end to this.

Because "beginning" means "there is nothing before it", and "end" means "there is nothing after that", but everything continues to exist and will not be lost.

In other words, everything does not have a beginning or an end, and it will continue to exist forever.

If you do not feel it is the truth, it is probably due to a misunderstanding of some sort.

From experience, we are aware that most things have beginnings and ends.

However, even if this is mostly true, it is not the absolute truth every single time.

We should also be aware of the fact that there is no beginning or an end in the existence of "everything".

In current cosmological theory, the singular point, and singular region of the Big Bang is said to be the beginning of the universe.

On first impression, this cosmological principle seems inconsistent with the truth: "There is neither a beginning nor an end in the existence of "everything".

However, it does not contradict itself if we consider the following:

If it is the truth that Big Bang is the beginning of a part of "everything", there will be no contradiction between my opinion and the Big Bang theory.

Going back from the present to the past, it is said that the singular region of the Big Bang is the place where time and space were created, but I think that it is only the beginning of a part of "the whole universe that exists forever".

Because, the idea of a singular point and a singular area is supposing the "null" around it and "everything" I am talking about also includes the "null".

Infinitely and everlasting "everything" has neither a beginning nor an end. I think that the big bang has taken place in part of it, and its part of time and space has started.

––––––

* "All except" and "Restrict all" are errors in words.

* Illusion are not null, they are false perceptions (i.e. some existence.)

* The word "absolutely null" is strictly a word error. It is impossible.

2: About Eternity: "Everything" Exists Eternally

The whole universe (Everything) exists. I proved the fact in my other essay, deducing from the truth "I think, therefore I am".

To exist means that it inevitably exists at a certain moment of time. Therefore, "Everything" exists at least at a moment of time.

Now, let's think about the nature of everything. "Everything" will never vanish as a whole because it includes the close-to-absolute null. For example, if the space would have changed into the close-to-absolute null, the space would exist as the close-to-absolute null. Since we are correctly perceiving (so it truly exists as it is) "so-called nothing (the close-to-absolute null)" by logic, "everything" as a whole will never vanish. Even if it might change or differ in some parts, it never vanishes as a whole. It means that "Everything" exists eternally.

Therefore "Everything" exists eternally.

The next point to consider is that time might not last forever.

We know from our experience that the world we are living is filled with changes. The fact that there are changes means that time (the difference and between moments) exists.

Then, what if everything stopped changing? It would mean the end of changes between moments. In that sense, everything exists eternally while time could end.

But we don't have to be in despair thinking that our world could vanish, because everything will never vanish. Thus, we can have real hope.

We can also have the confident hope that even if the world we live in ceases to exist, somewhere out there in the whole universe there will be other worlds rich in change and will continue to exist, because the whole universe is infinite in its expanse.

If something great has created the universe, we might go beyond time within an extent of the creation of the universe.

we can write a computer program. In the program, we can go beyond, stop, go back, go forward, copy, delete, fast forward, and delay time.

Then, what if "Something Great" has written the program of the universe?

In that case, we can say it is possible for us to go beyond the time in the universe.

(I consider that the time is the difference between the moment or a continuation of the moment. I disregard the sequence. We feel the time has sequence from our daily experience , But I think it is not always applicable.)

3: The Whole Universe is Infinite, Eternal and Transcends Time.

I will present the main thesis of this book hereafter in "3".

-

3-1: Abstract

-

This article presents the following truth: "the whole universe" (the whole world) is infinite, eternal, and transcends time.

The Big Bang would have happened in some part of "the whole universe."

To prove the truth, I apply logical thinking and Descartes's quote, "I think, therefore I am."

The main conclusions are as follows:

"The whole world" substantially equals "the whole universe" or "self and everything else."

"The whole world" ever lasts because it will neither become "absolute nil" nor vanish.

"The whole world" has neither a beginning nor an end, thus it transcends time.

It is eternal.

The truth I present may also suggest that we should spread happiness when we wish for peace.

That is because it is the universal value as we all share the same life.

We should achieve Joyous Coexistence in our ordinary life.

The word "whole" implies "holiness." I feel a mystery in the nature of the existence of "the whole universe."

-

3-2: Section I

-

"The whole world" equates to "the whole universe" or "self and everything else", which is infinite, existent and everlasting.

Firstly, that is because there is always the beyond in the universe so that it is infinite. (1)

Secondly, this is because the reality of existence of "self and everything else" can be proven by Descartes's quotes "I think, therefore I am." (2)(3)

Thirdly, this is because "the whole world" will never vanish so that it everlasts.

Since there is no "absolute nil" and even if "the whole universe" would be changed into so-called "nil", I regard it as still existing (I consider so-called "nil" is a kind of existence). (4)

Thus, "the whole world" is infinite, really existing and everlasting.

———————————————————————-

(1) If you travel the universe on a spaceship, there is no border.

This is because the beyond inevitably exists.

Even if the universe is surrounded by "nil", I argue the nil is also the beyond. ("the whole world" includes "the beyond")

(2)The fact that "If a subject does not exist, it cannot think" proves the existence of self and territory where the self must exist at or within (let B stand for the territory hereafter).

It also proves the existence of territory other than B.

(3) When one assumes B, one defines a territory other than B at the same time (Refer to Set in mathematics).

(4) Premise: ①If a thing is perceived correctly and given a proper name, I regard it as existing (As we give "nil" the name).

This is somewhat contradictory, the limit of my word power as of now. ② "Absolute nil" is a word error, that does not make sense.

3-3: Section II

"The whole world" (the whole universe) is eternal, because it has neither a beginning nor an end, and it transcends time.

Firstly, this is because "the whole world" is everlasting.

This fact seems to contradict "The Finite Universe Theory"-—the theory that insists there are limits to time and the universe and denies its eternity.

In the theory, a beginning means that there is nothing before it, and an end means that there is nothing after that.

However, "the whole world" always exists, and it means that it is everlasting. Therefore, it is eternal. (5)

Secondly, this is because, to everlast means to transcend time.

It is because, although time is the difference between two moments, or a continuation of the moment, to everlast also means to consistently and continuously exist at any moment of time. (6)

Thus, "the whole world" is eternal.

----

(5) From our subjective experience, we are aware that most things have beginnings and ends.

However, even if this is mostly true, it is not the absolute truth every single time.

We should also be aware of the fact that there is neither a beginning nor an end in the existence of "the whole world."

(6) I consider that time is the difference between two moments, or a continuation of the moment. I disregard its sequence.

We feel time is chronological from our daily experience, but I do not believe this is always true.

This is because, if time is like a computer program, it could be possible for us to transcend time.

-

3-4: Section III

-

In conclusion, as I mentioned above, "the whole world" is infinite, eternal, and transcends time, therefore, we should spread "happiness" which could be the best value in "the whole universe" without sticking to specific ideologies like ethnicities, religions, thoughts and nations. Most of those must include so many valuable concepts and should be highly respected. The examples are words of Jesus, like "love your enemy" and "love one another." (Jesus is the mutual holy one among Christians, Jews and Muslims)

The value of happiness is universal because we all are part of the same life. We should achieve Joyous Coexistence in our ordinary life.

The word "whole" implies "holiness." I feel a mystery in the nature of the existence of "the whole universe."

-

3-5: Supplementing articles

-

I will add the following two short articles to the above-mentioned main thesis.

3-5-a: "The whole world" and the Big Bang Theory

In current cosmological theory, the singular point, and singular region of the Big Bang is said to be the beginning of the universe.

On first impression, this cosmological principle seems inconsistent with the fact: "There is neither a beginning nor an end in the existence of "the whole world."

However, it does not contradict itself if we consider the following: If it is the truth that Big Bang is the beginning of a part of "the whole world", there will be no contradiction between my opinion and the Big Bang theory.

Going back from the present to the past, it is said that the singular region of the Big Bang is the place where time and space were created, but I think that it is only the beginning of a part of "the whole universe" that exists forever.

That is because that the idea of a singular point and a singular region is supposing the "nil" around it and "the whole world" I am talking about also includes the "nil."

Infinite and everlasting "the whole world" has neither a beginning nor an end.

I think that the big bang has taken place in part of it, and the part of time and space has started in it.

3-5-b: Something Great and the Program of the Universe

Something Great may have written the program of the universe.

If so, we can say it (the creation) might be imperfect.

Let us take it more positively however.

The program must be running almost perfectly in the whole universe, and some tragedies in our life have occurred by some minor errors in the vast program and they can be modified.

By the way, in the program, we can go beyond, stop, go back through, go forward through, copy, delete, make fast, and delay the time.

Then suppose Something Great has written the program of the universe.

In that case, we probably go beyond the time in the universe.

Supplementary Explanation:

A phenomenon that space arises from nowhere might be like one that would happen when someone just started to run a computer game program in which there is space (world).

4: Dialogue with the finite theory of the universe, etc.

<(1)>Dialogue with the View that the Universe has an End

—-

According to the book titled "*Is there an 'End' to the Universe?*" (*1), the latest scientific views suggest that the universe has a finite end.

<Quotation> Our universe, which originated from the chaos of the Big Bang within the vast emptiness of the Mother Universe, will eventually reach a state of eternal silence known as the Big Whimper, around the 100th power of the cosmic year. <Unquote> (p.13, translated by me)

The author of the book, Nobuo Yoshida, does not explicitly state whether the expanse of this Mother Universe is finite or infinite.

However,

Even if the expanse of the Mother Universe is finite, assuming the infinity and eternity of the "entire" universe, it does not necessarily imply the existence of only one Mother Universe. Therefore, even if one Mother Universe were to become "silent" in a "Big Whimper," other Big Bangs could occur in other Mother Universes. By assuming the possibility of other Big Bangs, the 'entire' Universe would not come to an end as a whole. Only a small part of the infinite universe may come to an end. Therefore, the infinite 'entire' universe as a whole will not come to an end.

Conversely, if we assume that the expanse of the Mother Universe is infinite, even if "our universe" were to experience the eternal silence known as the Big Whimper, there might still be other universes within the infinite whole, or new ones might be born.

However, according to the author of the book, while the possibility of countless universes being born can be suggested, mere ideas are insufficient. No leading theory has been developed yet to convincingly support this notion and gain the consensus of many physicists. he also emphasize that without a concrete, computable model, a theory cannot be regarded as a serious argument in the field of physics.

Regardless of whether the Mother Universe is finite or infinite, based on the aforementioned reasoning, the "entire" universe would logically avoid complete "silence." Only a small portion of the infinite universe may come to an end. Therefore, the infinite "entire" universe as a whole will not come to an end.

(*1) Nobuo Yoshida. *Uchu Ni Owari Wa Arunoka*: [*Is There an End of the Universe? - From its birth to '10 hundredth power years' later, as depicted by the latest cosmology*]. Kodansha, 2019.

—-

<(2)> Dialogue with the Idea That the Universe Is Finite in Time

—-

The "whole" universe is considered to be infinite in time (*). While the notion of the universe having beginnings in space and time may appear to contradict infinite time, this apparent contradiction can be reconciled through a multi-layered time model. According to this model, the finite universe that originated with the Big Bang is seen as a part of an infinite "whole" universe.

First, (1) I will explain that the whole universe is infinite in time, and then (2) I will explain the multilayered time model.

(1) To simplify the conclusion, the whole universe encompasses "nil" (Close-to-absolute nil I) with infinite spatial extent. As it remains unchanged and continues to exist, it endures infinitely in time. Therefore, the whole universe exists infinitely in time.

The explanation of the whole universe as a premise is as follows:

In essence, the whole universe consists of both "nil" and "thing." "Nil" can be further categorized into "Close-to-absolute nil" and "relative nil," both of which are perceptible entities and thus part of "existence."

In other words, the whole universe = everything comprises "Close-to-absolute nil I" + "Close-to-absolute nil II" + "thing" (including "relative nil = relative thing").

('Close-to-absolute nil I' refers to that which can solely be recognized through words and possesses spatial expanse.)

("Close-to-absolute nil II" refers to the subset of nil that can solely be recognized through words and lacks expanse.)

('Relative nil' is synonymous with 'relative thing' and denotes entities imperceptible by certain beings through their senses alone. For instance, ultraviolet light is imperceptible to humans through their five senses, yet it exists. (However, it can be detected through measuring instruments, etc.))

("Thing" encompasses entities whose existence is perceptible, excluding for convenience "Close-to-absolute nil I and II." It includes both "relative nil" and "relative thing.")

(Close-to-absolute nil I and II are considered existences since they can be perceived through words.)

Now, let us delve into the explanation of 'Close-to-absolute nil I' and its infinite existence in time.

Close-to-absolute nil I represents 'nil that can only be grasped through words and possesses spatial expanse.' Its spatial expanse stems from being devoid of all 'things' within space. ('Thing' encompasses not only objects but also phenomena such as radio waves, light, smells, and energy.) Therefore, it does not encompass any 'thing' whatsoever. Even if we assume the expansion, contraction, or distortion of space, we require a field (space) capable of accommodating and encompassing such transformations. In essence, Close-to-absolute nil I lacks any transformative elements and thus remains unchanged. From another perspective, it can be considered a foundational field in which "things" can be placed, as all "things" have been eliminated from it in advance. Its perpetual state of non-change implies that it continues to exist unaltered and never disappears, hence rendering it infinite in time.

The explanation for the infinite spatial expanse of Close-to-absolute nil is as follows:

When we consider a certain territory, we implicitly acknowledge the existence of something beyond that territory. Infinite space encompasses both the inside and outside of a given territory. 'Close-to-absolute nil I' exists within and beyond that territory, with an infinite expanse that extends inward and outward (this is because all 'things' have been removed from within and outside the territory, respectively). Therefore, if we designate a certain territory as the universe, the entire universe, encompassing the inside and outside of that territory, extends infinitely, and 'Close-to-absolute nil I' also extends infinitely.

As a result, 'Close-to-absolute nil I', serving as the backdrop for the existence of 'things', spreads infinitely like an expansive background, existing in infinite time.

Furthermore, this demonstrates that the notion of a 'temporal beginning and end of the whole universe (everything)' is a linguistic error. It reveals its impossibility.

(2) Multi-Layered Time Model:

In the limited universe that originated with the Big Bang, the concept of a beginning of time is assumed. (*)

This assumption aligns with the temporal infinity of the whole universe in the multilayered time model.

In other words, we can postulate the beginning of a particular layer of time in relation to the spatial and temporal infinity of 'Close-to-absolute nil I'.

For instance, by using a stopwatch and varying the speed of time progression (faster or slower than usual) for clarity, we can observe the existence of multiple time layers within a single time. Thus, within an infinite time, there can be additional time layers, including their respective beginnings and ends. Time can exhibit a multilayered nature.

Moreover, since the whole universe is infinite and eternal, the occurrence of the Big Bang does not necessarily represent a singular event. It is worth considering the possibility of the existence of multiple universes.

(*)My understanding and limitations regarding the definition of 'time' discussed here are as follows:

Time: The duration between moments.

Moment: A specific point in time when something 'exists'.

Point in time: A singular moment in time.

Duration: The difference, change, or continuation between moments.

Regarding time, anything beyond these aspects is currently not well understood for me. (as of May 2023).

(*) The Big Bang Theory suggests a beginning of time: Katsuhiko Sato. *Sotaisei-riron ni Okeru Jikan to Uchu no Tanjyo* [*The Birth of Time and the Universe in Relativity*]. Design of Space-Time. 2007. http://umdb.um.u-tokyo.ac.jp/DKankoub/Publish_db/ 2006jiku_design/satou.html, (ref 2023-5-2).

—-

<(3)> Dialogue with the Finite Theory of the Universe Deduced from a Three-Dimensional Torus Structure

—-

A three-dimensional torus structure is formed by connecting a two-dimensional torus (a ring-shaped surface resembling a doughnut) to the top, bottom, left, and right sides of a rectangle, and it is said that the front and rear sides are also connected. Unlike the two-dimensional case, where the distant edges are connected, the distant surfaces in a three-dimensional torus are connected, resulting in a continuous world where going straight in any direction leads to circling back around.

It is important to note that the analogy of a doughnut shape is only used to illustrate the concept, as the structure itself has no curvature. In other words, space is connected to space without any bending or warping. (Reference: Takahiko Matsubara, *Uchu ha Mugen ka Yu-Genka* [*Is the Universe Infinite or Finite?*], Kobunsha 2019)

Now, there is a notion that the universe has such a three-dimensional torus structure and that its expanse is finite.

However, even if the region of the universe we inhabit possesses a three-dimensional torus structure, the "whole" universe remains infinite. Alternatively, the proposition that the "whole" universe is a three-dimensional torus structure and therefore finite is incorrect.

Firstly, even if we assume that a three-dimensional torus structure is a physically possible finite structure, it must still be encompassed within an infinite space defined by the XYZ axes. This implies that the entire universe is infinite. In other words, being a finite three-dimensional structure, it would exist within the infinite expanse of the XYZ axes, and conversely, the XYZ axes would extend beyond the boundaries of the three-dimensional torus structure and further.

Furthermore, the three-dimensional torus structure is described as a loop-like structure without bends, meaning it is connected without any bending in the up-down, left-right, and front-back directions. This loop-like nature restricts one from surpassing a certain point, resulting in continuous circling. In other words, even if one were to travel "straight ahead" within this structure, they would be unable to surpass a particular point due to the looping effect, endlessly circling without reaching a new destination. Such a loop structure is conceivable if our universe is programmed in such a way. However, even if the universe we inhabit follows a three-dimensional torus structure with this looping characteristic, the whole universe remains infinite because it encompasses areas beyond the looping point, including the regions "beyond" its structural expanse, even if they are considered "Nil."

By the way, since the whole universe is infinite, it is plausible that multiple universes with three-dimensional torus structures and multiple universes without such structures exist concurrently.

—-

<(4)> Dialogue with the View that "Space is a Physical Entity, Distinct from 'Nil,'" or with the View that "Nil has No Spatial Expanse"

—-

I acknowledge the existence of 'nothing' (referred to as 'Close-to-absolute nil II') that lacks spatial expanse within the realm of language. However, there is also 'Close-to-absolute nil' (referred to as 'Close-to-absolute nil I') that encompasses a spatial expanse by removing not only objects but also all 'things' such as waves, light, and energy from space. This 'Close-to-absolute nil I' must exist as a distinct entity alongside 'Close-to-absolute nil II'. (Even if we consider the possibility of space being physically distorted, there must be a background that allows and encompasses this distortion, and 'Close-to-absolute nil I' serves as that background. Thus, space can exhibit a multi-layered nature. In essence, a layer that remains unchanged as a background can coexist with one or more layers that undergo change.)

In other words, it is plausible to posit that space, with its physical characteristics, extends with 'Close-to-absolute nil I' as its background, thereby creating a multi-layered structure. If space lacks physical characteristics and does not contain any 'thing', it is equivalent to 'Close-to-absolute nil I'.

An example that illustrates the concept of space being multi-layered can be found in the realm of virtual reality. The virtual space within a computer exists as a singular reality, while the world we inhabit serves as the background for that virtual space.

Another example (although it may not fully support the claim) is that of a container used in an experiment. If there is a distinction between the space within the container and the space in the laboratory, it could indicate a multi-layered structure of space. In this case, the space within the container and the space in the laboratory can coexist in multiple layers, with only the space within the container undergoing physical changes while the other remains unchanged.

In summary, there exist two forms of 'nil': those with spatial expanse and those without. It is a cognitive error to assume that the absence of space in a part of 'nil' represents the entirety of 'nil'. There exists a type of 'nil' that encompasses spatial expanse, which coexists with 'Close-to-absolute nil II' that lacks spatial expanse.

—-

<(5)> Dialogue with the Finite Theory of the Universe Based on Spatial Curvature

—-

The finite theory of the universe based on spatial curvature is described in Takahiko Matsubara's book "*Uchu ha Mugen ka Yu-Genka* [*Is the Universe Infinite or Finite?*]" (Kobunsha, 2019).

According to Matsubara's explanation:

"If it can be proven that spatial curvature is uniformly positive throughout the actual universe, then we can conclude that the universe is finitely closed. In such a case, the question of whether the universe extends indefinitely would be resolved. If you were to travel in a straight line within this universe, you would eventually return to your starting point, and space would not extend indefinitely." (Translation by me)

However, even if we assume that space is curved, traveling straight ahead would take us through that closed space and beyond, extending infinitely in any direction.

This can be understood by examining the misperception that arises when we "travel straight" on Earth. Normally, if we walk straight on Earth's surface, we eventually return to our starting point due to the Earth's spherical shape. This creates the misconception that if we were to travel straight "truly," we would return to our original location.

The following explanation clarifies the nature of this misconception:

In reality, we are not traveling straight. When we walk straight on Earth's surface, we are actually traveling "in accordance with the curvature of the Earth," rather than in a truly straight line.

If we were to travel truly straight, we would gradually depart from the Earth's surface and eventually travel through space.

Similarly, if we were to travel "truly" straight through space, we would not follow the curvature of space and thus continue beyond the confines of the closed space into an infinite expanse.

The question of whether we can travel without following the bends in space, assuming the existence of such bends, and the question of whether there is a space beyond where we can travel without following the bends are two distinct issues.

In conclusion, even if the universe in which we reside is finite and closed in terms of spatial dimensions, there exists an infinite expansion beyond its boundaries.

—-

<(6)> Dialogue with the Finite Nature of Cosmic Space: Perspectives on Expansion and Contraction

—-

One theory regarding the finiteness of the universe, based on the Big Bang theory and other related theories, suggests that cosmic space is finite and experiences both expansion and contraction.

In this view, the limit of the universe's expansion is determined by the "critical mass" of matter. If the amount of matter exceeds the critical mass, a phenomenon opposite to the Big Bang occurs, leading to the universe contracting back to a singular point. On the other hand, if the amount of matter is below the critical mass, the universe's expansion continues indefinitely (Toshifumi Futamase, "*Zukai Zatsugaku Uchu-ron*" [*Illustrated Miscellaneous Cosmology*], Natsumesha, 2002, p. 100).

However, whether space expands or contracts, the whole universe exists as a background, encompassing "nothingness" or "nil", and because it includes "nothingness", its expanse is infinite. (For further exploration of 'nil' and the infinity of the whole universe, please refer to 'About Infinity' [https://www.jcstation.info/e.infinity.htm]).

It is important to emphasize that the universe in which we reside is merely a part of the whole universe. Even if it were to reach an endpoint through a contraction known as the Big Crunch, other universes must exist and persist somewhere within the infinite expanse of the whole universe.

(Ⅱ) About Happiness

I will discuss happiness in this part.

1: What Is Happiness? (Dictionary definition of "happiness.")

To put it simply, "happiness" varies among countries and cultures, but I will introduce you to a good example of the definition later.

The definition of happiness is not all the same in dictionaries. The word translated as "happiness" has some similarities in meaning and content in different languages in different countries, but also some differences. In other words, it is difficult to compare the definition of happiness in the world. It is a problem which inevitably accompanies translation. The choice of words is considered only to be equivalent and must be arbitrary, even by authorities. Even in the same language, the definition of happiness may have its historical change.

> Such as, worldwide, Happiness (the equivalent term) is frequently defined as "good fortune" and "favorable external environment," but in American language it has been replaced by a definition that focuses on a "favorable internal sense state. (A comparison of 30 languages shows that 24/30 = 80% of the words considered "good luck" as an equivalent for "happiness").

(From *Concepts of Happiness Across Time and Culture*)

By the way, there might be a substance of "happiness" somewhere that has not yet been well defined in various languages. Or perhaps we expect too much from the word "happiness" or have illusions about it.

Under these circumstances, I would like to give a definition of "happiness" from Oxford English Dictionary as an example to later discuss how to be happy. According to the dictionary, "happiness" means: The quality or condition of being happy. And this is divided into three meanings: 1.(a) Good fortune or good luck in life generally or in a particular affair; success, prosperity. Now rare. (b) An instance or cause of good fortune.; 2. (a)The state of pleasurable contentment of mind; deep pleasure in or contentment with one's circumstances. (b) An instance or source of pleasure or contentment. 3. Successful or felicitous aptitude, fitness, suitability, or appropriateness; felicity. Also: an instance of this. Now rare.

The point is that we should consider the above definition to be good enough and not to expect too much from the word "happiness" or to add or subtract meaning. Otherwise, we would be at a loss in grasping what it is and would lose track of it.

This is because this definition not only covers most of the meanings in a simple, clear, broad and rich manner but, also because it partly is a subjective definition of fullness, so there will be no shortage of the meaning. That is, it fully covers the aspect that people have different minds and what makes them fulfilled.

By the way, I consider the above-mentioned definition could be focused into two meanings: 1.To be fortunate. 2. To be in a pleasurable contentment.

This focus will be used in considering (2) How Can You Be Happy?

2: How Can You Be Happy?

So how can you be happy? How can you fill your heart?

For individuals, I believe that this can be achieved by (a) satisfying your needs, (b) to attain a contented mind, (c) adjusting your needs (to some extent), and (d) thinking in a way that makes you happy. (e) to be fortunate

This is explained below.

- - -

(a) Satisfying your needs

In a narrow sense, the solution to the problem of how to be happy is: To satisfy all your needs.

According to psychologist Maslow's theory, people have five levels of needs in common. These are physiological, safety, social (i.e., love or belonging needs), esteem, and self-actualization. It depends on the person's situation as to which needs they have.

Physiological needs are the desire for food, water, oxygen, etc. While social needs are the longing for friends, lovers, and a place in a group. Moreover, the need for esteem is the desire for self-esteem, respect from others, status, and a sense of accomplishment. Lastly, the need for self-actualization is the desire to find oneself and to grow more. This desire for self-actualization is said to arise only when all other needs have been sufficiently satisfied.

Therefore, you should satisfy the above-mentioned five needs; you should eat well, live well. get friends or partner, get esteem, and actualize yourself, etc.

- - -

Here are some tips:

You can try Google search to find the answers to the questions such as how to have friends, loved ones, and be accepted by everyone, or how to actualize yourself. When it comes to the desire to "find oneself," you should recognize the difference between your goals and your current situation, then think about how you can fill the gap. You can also seek some help (such as reading self-help books or getting advice from a good friend or wise people).

- - -

One more comment on satisfying the desire for self-actualization:

I think that human beings are basically mere life forms (creatures) who seek pleasure and keep away from pain, but we are also the ones who set up ideals and strive towards them. Therefore, I think that both loving yourself as you are and realizing your ideal self are necessary to fulfill your desire for self-actualization.

- - -

(b) To attain a contented mind (Be satisfied with one's lot in life)

To sense that you have "enough" should also give you a sense of happiness.

It is also true that there is no limit to our desires, since we usually seek more and deeper pleasure, but there are also desires that we can control, which can increase or decrease at the level of our consciousness.

We have a certain amount of control over what we direct our attention to. When you are aware of what you lack, what you want and what you do, you will have a feeling of that something is missing, which will result to the secretion of brain chemicals related to it. In other words, a sense of actual scarcity would arise from consciousness.

If this is the case, we can assume that if we focus on what we are satisfied with, brain chemicals related to this will be secreted, and we will feel satisfied to some extent. (The exception to this is that attention and awareness may turn on to what we want automatically. (In that case, where we may feel impulsive scarcity.)

(However, I still don't know the solution to the problem: Once you know first-class, you won't be satisfied with second- or third-class. I wonder what we should do in this case.)

- - -

(c) Adjustment of desire (d) Thinking in a way that makes you happy

By the way, I reckon that people's desires seem to be adjustable to some extent by changing the way they think.

In other words, when we change our mindset, we change our perception of the situation. When the perception of a situation changes, the brain chemicals secreted also change. I conclude that if the brain chemicals that are secreted change, there should be an increase or decrease in desire, thus an increase or decrease in sense of happiness.

For example, if your anxiety increased about the same event or fact, worrying that "What if X (a bad thing) would happen?", you will feel anxious and actually feel unhappy. Instead, if you start to positively think about it, in a way such as "anxiety is a warning signal" and "let's not care much about it after taking as much countermeasures as you can" you will feel less anxious and feel happier as dopamine and other positive brain chemicals are secreted from your positive attitude. As such, whether you feel happy or not, partly depends on exactly how you think of the situation.

By the way, since there are many kinds of desires, there must be many ways to be happy.

For example, people should have different ways of satisfying physiological needs such as their appetite, etc.

Each person is different even in choosing "hobbies," which is supposed to belong to a higher level of desire.

It is also true that when it comes to the desire for self-realization, or to "become who you want to be," it should be different for each person.

- - -

(e) To be fortunate

Luck(fortune) is often perceived as an element of chance. However, it may be possible to improve it by your own willpower to some extent. Firstly because, causality is generally true, from the mechanistic view of the world. In other words, if you do good deeds, good results usually come back to you. In other words, if you do good deeds, you will be lucky.

Secondly because, we can expect to have good luck by believing that we are lucky.

This is because humans have a cognitive ability called selective attention. In other words, we are less likely to recognize what we do not consciously pay attention to, but we are more likely to recognize what we do pay close attention to. In other words, when you consider that you are lucky, you are more likely to find good fortune, or in other words, you will be lucky.

Here, I will use psychiatrist Shion Kabasawa's November 2020 seminar, "How to Be Happy," as support for this essay. It is a way to be happy in terms of concrete actions.

Simply put, the way to be happy is to keep your mind and body fit, strengthen your connections, and work hard.

How to Be Happy: a Seminar by Zion Kabasawa, a Psychiatrist.

Conclusions of the seminar:

- - -

Things you need to do:
1. Be healthy and keep sound mind
2. Strengthen the ties with your loved ones: Avoid isolation.
3. Work vividly

- - -

To do (more concretely):
1. Sleep well, exercise well, do morning walk.
2. Keep a good relationship with your family members, friends, pets. Do volunteer work, or charity,
3. Do the PDCA Cycle (Plan, Do, Check Action) and give reward to yourself when you accomplish each plan. Praise or psyche up yourself.
4. Ask yourself "When do you feel the happiest?" then try to recreate those moments.
5. Find three happy things every day and write them in a three-line short happy diary at night, and go to bed feeling happy.

- - -

Because:

- - -

By doing the things above, we feel happy because happy brain chemicals mainly Serotonin, Oxytocin, Dopamine will be secreted adequately, and will be balanced.

3: What Is Eternal Happiness?

So, what is "eternal happiness"?

It is "a state of being content transcending life and death."

I think eternal happiness is not completely reachable but able to get close to it, because nothing is unchanging but also not are our states of mind.

However, it may be possible to get very close to it, if we can also live in the spiritual world, and/or the digital world. Because if we can transcend the limitations of life; birth, sickness, aging, and death, we may be able to overcome the sufferings accompanying them.

Scientifically speaking, living as an organism should involve suffering or unhappiness. For example, people feel uncomfortable if they get hungry, feel pain when they get sick. Aging will bring decline, and death is usually accompanied by the fear and pain of disappearance.

However, if we can also live in the spiritual world and/or digital world, we can expect that such life won't accompany the above-mentioned sufferings.

In this life, a state close to eternal happiness should be "peace = joyous coexistence, immortality, and playful life. We should not totally deny this life completely in face of the above-mentioned sufferings. This is because total denial of this life is destructive, and although there should be sufferings in this life, we are able to reduce them and rather improve this life through medical and social improvements.

That is, through genetic engineering and advanced medical treatment, we will be able to achieve permanent youth, overcome many diseases, and transform our lives as living beings into something less painful and more satisfactory for a longer period of time. In addition, we will be able to be content individually when we establish a social system that guarantees a variety of ways of life.

To achieve this, the first thing we require would be a society that is free from "fear and lack" and protects lives through providing a high level of security, freedom, and safety.

Respect for other lives is also important.

In this life, I believe that not only humans but also other living beings should be happy. However, in this world, there are realities such as the struggle between the weak and the strong, fight against fierce animals, pathogens, and pests. I can't think of a way to fully achieve eternal happiness in the reality. Nevertheless, it is possible to minimize victims through introducing segregation, developing genetic engineering, and eating meat substitutes or artificially cultured meat.

4: What Is a Happy Society?

Simply put, it is a society where everyone is happy (being content). The key is "everyone." If some people are suffering and grieving on a constant basis, we could not consider it a happy society. A society can be a happy society even though happiness may differ from person to person.

So, what should be the goal for a happy society?

If the goal is to realize a happy society, then the goal should equal to have the content of heart for each person in the society (group of people). Therefore, satisfying our common needs and desires as living beings and satisfying the needs and desires of our diverse minds should be the goals of each individual and policies. Specifically, the goals should be to provide a wide variety of common items, such as water, food, education, security, and entertainment, to meet the diverse needs of each individual.

when we consider the above-mentioned things in reality, it must be effective to utilize various sources of knowledge such as biology, sociology, economics, psychology, and religious studies.

5: Conclusions

Although it is difficult to compare the definition of Happiness worldwide due to translation problem, etc., I chose the Oxford dictionary as an example to discuss how to be happy and the points of the definition are: good luck; state of pleasurable contentment of mind; deep pleasure in or contentment with one's circumstances.

In a happy society where everyone is happy, one can be as such by fulfilling all his/her needs, to sense that having enough, adjusting desires, getting luck, and thinking in the way that makes you happy.

Happiness is not a delusion or mere dream but a reality we can get or live in. eternal happiness can be closely achievable if we transcend life and death or can be close to it even in this life by social and medical development.

We should live in happiness now, while savoring the gifts from Something Great, we should try to improve our society for the better towards eternal happiness.

————

(References)

—————-

*Saul Mcleod, PhD. "*Maslow's Hierarchy of Needs*". Simply Psychology. 2022-04-22.   https://www.simplypsychology.org/maslow.html, (accessed 2023-2-7).

*Hideyoshi. *Shinrigaku No Jikken De Manabu Ko-Un No Mitsuke-Kata* [The way to find how to be fortunate, a lesson learned from a psychology experiment]

https://www.lucky-rookie.com/gorilla/, (accessed 2023-2-9)

*Tokio Honda *Shiawase Ni Naru Kangae kata* [The Way of Thinking that Makes You Happy], Alpha Police 2007

**Dictionary of Psychology*, Maruzen

*Shion Kabasawa *No Wo Saitekika Sureba Nouryoku Wa NIbai Ni Naru* [If you optimize your brain, you can double your ability], Bunkyosha 2016

*Shigehiro Oishi, Jesse Graham, Selin Kesebir and Iolanda Costa Galinha *"Concepts of Happiness Across Time and Culture"*◇SAGE Journals 2018.

(Ⅲ) Other essays and articles

I will present the following ones in this part.

—-

1: Fundamental Philosophy of Peace

2: For a Better World.(Let's avoid generating criminals)

3: Two Angels Who Will Save People from Hell: "Basic Income" and "Helicopter Money"

4: Today, a Child Is Dying of Hunger Every 5 Seconds

5: What do you say to my plan, "Save All Now and Forever"?

6: What Is the Main Cause of Poverty?

7: Letter to Avengers

8: Coexistence of Libertarianism and Everyone's Happiness

1: Fundamental Philosophy of Peace

Contents

1. Introduction

Even now, many armed conflicts are ongoing in the world. It is unusual for people living in developed countries like Japan to actually and deeply know the tragedies of wars or armed conflicts. Such people would only touch a fraction of the tragedies through news media from somewhere or by a short passage printed in history books at most. But we can imagine there were people in the wars or armed conflicts who watch their houses burned in dismay, who shed sorrowful tears facing the corps of their beloved ones, and people who had everything taken away by the inferno of atomic bombs.

We can assume that the people, especially those in such tragedies, have sought to find a better society in the word "peace". Then, what is true peace?

In this essay, I show my answer to the question and present it as the Fundamental Philosophy of Peace.

2. What is Peace?

Firstly I compare several concepts that are translated into "peace" in several languages.

"Santih" (Sanskrit): The main meanings of this word are "spiritually fulfilled peace of mind", "inner tranquility", and "calmness of mind that could not be disturbed by pains or pleasures".

"Pax" (Latin): In addition to the meaning "peace" in English, this word also means "treaty".

"Mir" (Russian and Serb-Croat): In addition to the meaning "peace" in English, this word also means "world"

"Frieden"(German): In addition to the meaning of peace in English, this word is said to be etymologically derived from "frei"(freedom) and "freund" (friend). And its origin from Indo European word "pri" is considered to contain the meaning of "to love" and "to care"

. "Eirene"(Hellenic): The main meaning of this word is "a state of no war" and "a short term between wars" same as "Negative Peace".

"Shalom" (Hebrew): In addition to the meaning of peace in English, this word includes a lot of meanings in it; It primary means "wholeness" and "perfection", and the shade of meaning are "accomplishment", "completion" "maturity", "healthiness", "integrity", "community", "harmony", "calmness", "safeness", "happiness", "welfare", "friendship", "coordination", "success", and "prosperity".

"Salam"(Arabian origin): In addition to the meaning of peace in English, this word also means "healthiness" and "greeting". According to ""Koran"" this word also means "peace of this world and the other world"

"Heiwa" (Japanese): In addition to the meaning of peace in English, "hei" means "equality", " justice", " balanced states" and "harmony". "wa" means "softness". It is also said that "wa" means a state of people not striving and so eating grains, and that "a tender language like an appearance of ear of rice bending before the wind".

3. Semantic Classification of "Peace"

These concepts of "Peace" are semantically classified into four: Inner Peace, Outer Peace, Negative Peace and Positive Peace. Inner Peace means peace inside of man, i.e. peace of mind. Santih is classified into this. "Outer peace" means a peace in a social relationship. Negative Peace means "absence of war". Positive Peace means a state of society which is affluent, harmonious, safe, democratic, respecting human rights, promoting welfare, filled with arts, fairness, richness, etc. in addition to absence of war. (Mr. Gultung's definition of Positive Peace is different. For him, it means "integration of human society")

Positive Peace is conversely derived from Negative Peace, regarding the latter as state of no peace, since it might include incompletion such as state of oppression, wide gap between rich and poor, and a colonial oppression even if there were no actual wars happening during Negative Peace.

4. The Semantic Limitation of "Peace"

I cannot regard all of the above mentioned concepts of peace as complete ones, because sufferings and conflicts among human beings might exist in the state of them. For example, even if somebody could attain, "Santih", a peace of mind, it is possible that his neighbors might be starving or freezing. In a Negative Peace, such as "Eirene", bloodshed might exist. Even if a society was filled with a lot of products and arts, and if all of the existing human rights are respected, they cannot fulfill all needs of the people perfectly. For example, the Right to Sunshine did not exist until recently.

5. Substantial Aspects of Life

Since the world, to which human beings are destined to be located, is always changing and their physical shapes and conditions, along with psychological are changing from birth to death, it is impossible that human beings can reach the permanent peace of mind, although it can be attained to some degree.

Whatever discipline a person attains and even if a person reached a spiritual enlightenment through a strict discipline of Buddhism to fast until entering into death being away from his neighbors, It must be impossible and meaningless for us "living" in a society. Accordingly, the peace of mind of human beings could only be attained intermittently in the substantial life. In short, "Inner peace" is an intermittent one.

It must be the same as "Outer peace". In a relationship in a certain situation, people can act in a way that won't be unpleasant to each other almost perfectly, while sometimes they dispute contrary. In other words, in human life and its social relationship, we can assume two opposing extreme; permanent pleasure, and permanent pain.

6. Joyous Coexistence.

In light of the above representation, I am going to present the new concept of peace.

It is "Joyous Coexistence."

More particularly, it is a state in which people hold a permanent joy as an ideal where everyone permanently live a life full of joy and likewise in their relationships, then will be content with the status quo when they achieve a certain level of quality of life, and still, seek for a better condition progressively when they have an extra energy.

The reason why the ideal will finally be a state "to seek progressively" instead of "to achieve" is that we cannot achieve a permanent joy in reality.

I consider that this is a new concept of peace: Joyous Coexistence, which makes up for limitation of the previous ones.

Supplementary Explanation: The reason why I mentioned "become content with the status quo when they achieve a certain level (of quality of life)" is that seeking perfection excessively without affirming the status quo has a opposite effect to their joyfulness.

There are sayings like "the desire of perfection is the worst disease that can afflict the human mind."(Emerson) or "A contented mind is a perpetual feast."

Seeking perfection is a noble thing, however, doing it without affirming the status quo causes displeasure and dissatisfaction, and leads to unhappiness.

Perfectionism could be a cause of depression.

Therefore, it is also necessary to be satisfied with the status quo, and to seek for a better society after that.

However, we should not be satisfied with the status quo if we are in extremis such as an incessant and a terrible hunger or conflict.

It obviously has room for improvement, since then, we should "accept" reality in the firsthand.

In other words and generally speaking, we should not be immediately satisfied with the status quo while accepting reality until a certain level of needs in biological and social sense would be fulfilled, then achieving a certain level and taking the state as satisfactory, we should seek for the improvement.

Also, it may be important not to be aware of some problems or ideals.

Because just to know about them, even if it has nothing directly to do with someone, it could lead to his/her unhappiness in cases where he/she feels unpleasant or frustrated because he/she thinks he/she cannot solve, improve and cope with those problems, or achieve those ideals.

7. Possibility of Realization of Joyous Coexistence

It could be criticized saying that it is impossible to realize the Joyous Coexistence because of the selfishness of human beings.

Yes. It is impossible to "attain" Joyous Coexistence completely. As I have mentioned above, the goal of the ideal is not a thing to be attained but sought. Actually, because of the selfishness, we occasionally act being far away from Joyous Coexistence, in the dispute of insisting only one's own interests. For example, people sometimes favor conflict, even in the case where it damage others. I can show one typical example among many of them; A Short Account of the Destruction of the Indies by Bartolomé de las Casas. And even if some of us set it as a goal to realize the state where man-made products are affluent and hunger is overcome, not necessarily everybody hopes for that. But by the act of adjusting interests, even being selfish, we can realize Joyous Coexistence to some degree. For example, when we divide works, the self-interested activities could be correspondingly profitable for the others, although not perfectly: Egoism and altruism do not always contradict. Human beings have ability to adjust things and are able to narrow the gap of the egoism and altruism for its coordination. I consider these are very important for Joyous Coexistence. Selfishness should not necessarily be morally denied. Since if a person did not fulfill his self-interests at all, he would be unpleasant and Joyous Coexistence would not be realized. Contrary, if a person never does any good for others, i.e., if one's deeds were not profitable at all for others, Joyous Coexistence would never be realized.

So I consider it is important to adjust the balance between egoism and altruism for its coordination in order to realize Joyous Coexistence.

8. Love Your Enemy

It is also important to love even your enemies for Joyous Coexistence. The enemies of yours might be ones whom you hate and sometimes you might hurt. But through the deed of love (agape: unconditional love. It is different from neither "Eros" nor "affection"), the conflicting relationships in which people even hope the opponent's death could be changed for the better. By the act of loving our enemies, we will not miss the opportunities to understand our opponents and maintain the positive possibilities to alter the relationships for the better.

9. In Reality

Since the aspects of human relationships have wide variety and human beings are imperfect, the state of "Joyous Coexistence" in a reality is diverse, comparative and not perfectly attained. If I could indicate examples of Joyous Coexistence, they would be the ones that are much more closer to Joyous Coexistence compared with other states of relationships, since any state is not perfect. For example, if we could achieve a day when Jews and Palestinian could live together joyfully and respecting each other. This was the case in the days of Former Yugoslavia when different ethnic and religious groups had coexisted as good neighbors sharing various good things like food, smiles, etc. among them. This is an example of "Joyous coexistence". However in the present situation in Palestine, there are conflicts, many are killed by terrorist acts, and Palestinian's employment rights are badly restricted and their standard of living is poor

10. Conclusion

After comparing the several concepts that are translated into peace, I presented a new concept - "Joyous Coexistence".

Considering the substantial aspects of life, Joyous Coexistence is an idea in which the goal is not a thing to be attained but to be sought. However, it is gained by satisfying the status quo at some degrees.

To love your enemy is also important element in seeking the idea; in order to overcome a hostile relationship and change it into a friendly and joyful one.

It is clear that in light of this new concept, there still remains a lot of problems to be solved. I believe we can solve these remaining problems because we have already made a lot of dreams come true.

———-

(Reference)

————

* Yoshiaki Isaka, *Heiwa No Kadai To ShuKyo* [Issues of Peace and Religion]Kosei Shuppan, 1992

* *Kokusai Seijigaku Jiten* [Dictionary of International Politics] Tokyo Shoseki

*Heiwa-Gaku -Riron To Kadai -[Peace Studies: Its Theory and Challenges] Waseda University Press, 1983.

*Bartolomé de la Casas I, *Tratados de fray* [A Short Account of the Destruction of the Indies], Iwanami, 1976.

*The Early Lectures of Ralph Waldo Emerson, 1838-1842, Harvard University Press, 1972.

*Martin Luther king, Jr. Strength to love

*James pool and Suzanne Pool Who Financed Hitler : The Secret Funding of Hitler's Rise to Power 1919-1933. (The Dial Press, 1978)

2: For a Better World.(Let's avoid generating criminals)

As setting of the official interest rate of central banks eventually determine the (un)employment rate, I strongly suspect that some policies indirectly determine the crime rate. I also strongly suspect that some policies must be accompanying the invisible force that I believe to exist and affects people's mind, emotion, soul and heart.

So I invoke the better policies for a better world.

The official interest rate of central banks eventually determine the (un)employment rate. Although the interest rate does not directly effect the (un)employment rate, the fallout of the monetary policy effect the social conditions.

As such and therefore, I strongly suspect that some policies indirectly determine the crime rate. In other words, as policies are set goals, I suspect that criminal rate is also a set and an attained goal. Some of the statistical data must be, not a mere result but set-and-attained goals.

The crime rate must be set, to generate "tensions", in line of the saying "Aseptic society will perish" and "avoid perishing", that is, to maintain and improve the society. The "tensions" here means any kind of social unrest.

As an example, by experiencing the criminal incidents firsthand, the society will gain the response capabilities against other and similar criminal incidents. Furthermore, by the on-the-job training, the police are maintaining and improving their capability.

Thus, the society has been maintained and improved, just as a body gain the "immune strength".

However, we cannot totally approve the present way of maintaining and improving.

To know the reason, let's take a closer look at what "policies determine the crime rate" really mean, stepping into the dark.

That means the social policies indirectly make some people become criminals. Then when we become criminals? For example, let's think about schools. Generally speaking, as one of the social policies, modern public schools have been built and have contributed to make a stable society. But in the dark side, I strongly suspect they have indirectly generated the future criminals by forcibly tearing apart the heart of some students horribly enough to fall into the (so called) "evils", affecting them not only by the set school days, but also by the school days devastated behind the scenes by the invisible force that I believe to exist. The reason to generate criminals is to generate "tensions" in society.

However, without the tensions and the victims (victims are people who are fallen into the (so called ""evils")), we might fall into the terrible society as a whole.

So I invoke the better policies for a better world. We must not give up the better world, saying that the status quo is the best and we don't have any other choices. Since when we watch or feel the sufferings of the victims generated by some of social policies whether their effects have been direct, subsidiary or indirect (fallout), we definitely cannot say such policies are the best.

No more victims. No more breaking hearts. Much more love, healing and happiness for all.

For a better world now and forever.

3: Two Angels Who Will Save People from Hell: "Basic Income" and "Helicopter Money"

A young man from Japan visited a refugee camp in Afghanistan and saw Hell (*1).

People without arms, legs, and with burns all over their bodies, who must have fled from the battlefield were seen everywhere. The sight of them reaching out their hands to him, asking for one rupee (about two cents) as if they were hell's zombies, is lingering.

One of them, an elementary-school-aged girl who had lost both eyes, came staggering up to him in a distressed voice, saying, "One rupee, one rupee."

He could have given her a rupee or so. But he was too overwhelmed by the scene to do so.

In Rwanda, he spoke to a street child.

The child said, "The leftover food is good because it is a mixture of different kinds. When I get a lot, I sometimes use it to barter with other children.

They use the matches they get from bartering to make fire to survive the cold nights.

They can't wash or bathe. Their shirts and pants are full of holes. They spend their time under bridges, in the shadows of buildings, or in manholes and freight trains when it is cold.

The number of such street children in the world reaches about 100 million(*2).

According to an article on the website of Médecins Sans Frontières (*3), as of May 2021, about 3.5 to 5 million children under the age of five die annually from malnutrition, lowered immunity, and infectious diseases. According to a UNICEF article, as of 2018, 800 million people in the world are suffering from hunger (*4). Yesterday, some of they died, the same happened today and it can happen again tomorrow.

Now, let's turn our attention to Japan.

The Corona disaster has taken a heavy toll on women, especially those working in the hotel, restaurant, and service industries. (*5)

Just a few examples that can be relevant: a young woman searching for a way to sell her kidney on a computer in an Internet cafe where she can also sleep at a low cost, or a young woman wearing a fluffy, brownish, dirty mask and disappearing into Kabukicho(*6) saying, "I don't want to be on welfare.

I don't mean to assume that these things are hell in themselves, but they are like hell if they cost you so much.

To such hell, two angels will fall down from the sky.

Their names are "Basic Income" and "Helicopter Money." (*7)

Basic income is a plan to provide a certain amount of benefits to each citizen on a continuous basis.

Helicopter money is a real magic hammer.

It is a policy thats provides money directly to the people, just like a helicopter dispersing money from the sky. It is financed by the proceeds of money issuance.

The "proceeds of money issuance" means, in case of issuing a 100,000 yen gold coin, there will be a margin of 60.000 yen after deducting the cost which is 40,000 yen.

In case of a 10,000 yen bill, a margin of 9,980 yen is left after deducing the issue cost of 20 yen. This "helicopter money" can be used as part of the financial resources for basic income (*8).

By the way, helicopter money could bring about inflation if used badly, but it is a real magic hammer if used properly.

Having a national debt does not immediately lead to inflation. It can be used to secure financial resources at a level that does not lead to inflation.

For example, the world's outstanding government debt is about $90 trillion (*9), but the world is not experiencing inflation.

Japan, the U.S. and the EU each disbursed about $1 trillion for Corona-related reasons, but there was no inflation, this is another example.

I agree with the idea that these national debts should become perpetual bonds or can be written off in the end. The national debt is ultimately the debt of each citizen, and there are times when moderate repayment is necessary. Still, if no problem arises when the debt is written off, this equates the citizen's debt being dismissed.

The "hell" I mentioned earlier will change a lot with these two angels, "basic income" and "helicopter money." Many people will be free and escape hell.

Let's imagine this, if enough money is given to people in the hellish situations described above.

If there is a little more than one rupee, the refugees in Afghanistan can get food, clothes, etc. With more benefits, they may be able to receive medical care. The refugees mentioned above can sit around the table with their families with smiles on their faces. If they are sick or injured, they can be treated by doctors to heal their wounds and pains. I can imagine them searching for a new life with smiles and hope, not worrying about tomorrow's meal.

Street children will no longer have to worry about food. They can get clean clothes and a decent environment. Depending on how the helicopter money is used, slums could be redeveloped into decent living environments and children will be able to go to school. Children who have been oppressed on the streets will be able to smile, eat a full meal, play a lot, and sleep well.

What about in Japan?

For example, 70,000 yen (about 700 dollars) per month is provided unconditionally. Even the woman I mentioned before can continue her living and expect a chance for a better life without having to sell her kidney, without having to go on welfare, and without having to feel guilty about anything (*10).

What about the woman who was trying to work in Kabukicho? With a monthly benefit of 70,000 yen, for example, she won't be forced to work in the adult entertainment industry because she cannot think of anything else to do, but she will have the time and money to consider other opportunities.

In short, it would provide unconditional, regular income or an increase in income. Even if living on 70,000 yen a month is difficult, it will not only serve as a safety net that improves welfare, it also increases income and somehow eases poverty.

In addition, since many people will be able to choose jobs with more margin, we can expect improvements in poor working conditions. For example, in the hard positions that people will not be willing to do, we can expect their wages to rise and the work intensity and hours to be eased.

And the world will see an alleviation of hunger, poverty, and labor, and more happiness will descend upon the Earth.

By the way, the third angel, "reverse child allowance" may be necessary in some areas.

Reverse child allowance is a regular benefit paid to a woman based on the smaller number of children she bears. It has the effect on both controlling population size and promoting the protection of women. In other words, it is an angel that has the power to control overpopulation that may occur in some areas due to the global introduction of basic income.

Another important point is that basic income should not be implemented only in few countries, but should be implemented on a global scale with international cooperation. This is because, for example, if only one or few countries introduce basic income, issue a large amount of their own currency, and use the money to buy a large amount of goods from other countries, such as the U.S. and the EU, the currencies of those countries would plummet internationally.

Nevertheless, some of the determinants of the value of a country's currency may be the extent of its currency issuance and its relative position to other countries.

Therefore, countries such as Japan and the United Kingdom may be able to safely introduce basic income if they refrain from excessive currency issuance and if major countries such as the EU and the United States introduce basic income at the same time.

Let's Change Today and Tomorrow

We can do this by voting for political parties that promote basic income and lobbying the US and Japanese governments to work through the implementation of basic income with the UN and summits to reach a global scale.

Conclusion.

The introduction of helicopter money and basic income worldwide would alleviate poverty and hunger globally, this would help many people escape from hell. But first, let's vote for it.

- - - - - - - - - - - - -

(References)

—-

(*1) Stories of refugee camps in Afghanistan, stories of street children: *Honto No Hinkon No Hanashi Wo Shiyou* [Let's Talk About Real Poverty]Kota Ishii, Bungeishunju, 2019

(*2) Number of street children = 30-100 million:

https://www.unicef.or.jp/children/children_now/philippines/sek_ph02.html

(*3) From the website of Doctors Without Borders.

(*4) 800 million people are suffering from hunger:

UNICEF website

https://www.unicef.or.jp/news/2018/0151.html

(*5) Examples of women suffering in Japan:

[Women's poverty and suicides are skyrocketing! Real voices calling for politics and cries of 'Death to Japan] Shukan Josei PRIME, December 13, 2020

(*6) To work in Kabukicho implies to work in a sexual entertainment industry.

(*7) About basic income and helicopter money: *AI Jidai no Shin Basic Income Ron* [The New Basic Income Theory in the Age of AI] Tomohiro Inoue, Kobunsha Shinsho, 2018

(*8) Another part of the funding source for basic income should be tax revenue. Tomohiro Inoue, the author of the article mentioned in *6, envisions a two-stories basic income system using tax revenue and the helicopter money.

(*9) The world's government debt is about $90 trillion:

[Global Government Debt Equal to GDP, IMF 20-Year Forecast].Nikkei E-Edition, October 14, 2020.

(*10) About some of the problems with the current welfare system:

For example, people are advised to rely on family and relatives first, in some cases some of their property is confiscated, or if they earn income, their benefits are reduced accordingly.

AI Jidai no Shin Basic Income Ron [The New Basic Income Theory in the Age of AI] Tomohiro Inoue, Kobunsha Shinsho, 2018. *Rupo Seikatsu Hogo* [Reportage on Welfare] Ryoichi Honda, Chuko Shinsho, 2010.

- - - - - - - - - - - - -

4: Today, a Child Is Dying of Hunger Every 5 Seconds (*)

This means 14.000 deaths of children a day, 3.5 to 5 million deaths every year.

However, there is a concern that our world won't be sustainable because of overpopulation and expansion of consumption if we overcome the hunger and poverty.

Then, I set several relevant questions and have sent them to some people and organizations from whom we can expect positive answers.

—-

(A) Isn't it quite possible that our society will not be sustainable if overpopulation and expansion in consumption would occur when it becomes affluent for all people and we get over the poverty and hunger?

(B) If it is, what is the solution?

(C)What is an adequate world population where all people live an affluent life, setting living in Denmark, or the U.S., or other suitable countries as a standard?

—-

Although it must relate to the matters including natural resources depletion, environmental issues, development including space exploitation, and advance in technology, I think we need to seek the answer.

This is a quite serious problem in the real world where more than 20,000 people are dying of hunger day by day.

—-

I will post those answers in this page if I can get some.

→ I have got some positive answers

One possible solution is as follows:

(A)(B)By peacefully transitioning to an adequate population with the help of women's empowerment and other things, such as "Reverse child benefit(*1)"we can make our society sustainable even if all people enjoy an affluent life, overcoming the poverty and hunger.

Moreover, overcoming hunger and poverty; and educating poor people will lead to the expansion of consumption (economic growth,) namely, the leveling up of the world economy—these policies will be effective in protecting and fostering the current and potential human resources as both good labor forces and consumers.

- - - - - - -

(*1) "Reverse child benefit" is a benefit paid to a woman according to the number of children she has: the smaller the number of children, the more the benefit. This has a population control effect in heavily populated areas. It may sound rather rough, but I think it's a good idea.

(C)According to information from Wikipedia (as of June 4, 2021), the world's optimum population is 1.5 to 2 billion people (*2)

- - - - - - -

(*2)According to a study by Paul R. Ehrlich et al, the following were taken into consideration. The calculation seemed to involve:

*Adequate wealth and resources for all people
*Basic human rights for all

*Protection of cultural diversity

*Allowances for intellectual, artistic, and technological creativity

*Protection of biological diversity

My supplementary note: The calculation in his paper is based on the total amount of energy that can be consumed divided by the per capita consumption of developed countries (*3). It may be that the introduction of clean energy such as hydrogen has not yet been calculated.

(https://ja.wikipedia.org/
wiki/%E9%81%A9%E6%AD%A3%E4%BA%BA%E5%8F%A3)

(*3) The calculation was as follows.

The marginal energy consumption is 9TW (Terawatt = Terra is the 12th power of 10). The marginal energy consumption is 9TW (terawatt = terawatt is the 12th power of 10).

Realistically speaking, we need to take into account a 50% error, 6TW, additionally considering a 100% error, 4TW is the upper limit of energy consumption.

The average energy consumption of a developed country in the 1990s was 7.5 KW (kilowatts), but with modern technology, this can be reduced to 3 KW without compromising the standard of living.

6TW, 4.5TW divided by 3KW = total energy consumption per capita in developed countries = 2 billion, 1.5 billion

=the optimum world population = 1.5 to 2 billion

Reference: Gretchen C. Daily, Anne H. Ehrlich, and Paul R. Ehrlich. *Optimum Human Population Size* Archived 2017-08-17 at the Wayback Machine.. Population and Environment: A Journal of Interdisciplinary Studies Volume 15, Number 6, July 1994 01994 Human Sciences Press, Inc.

(*)The number of children under the age of 5 die of malnutrition, and immune deficiency or infectious diseases related to malnutrition amounts to 3.5 to 5 million every year. It means 1 death in every 5 seconds, 14.000 a day (From Web site of Doctors Without Borders) https://www.msf.or.jp/landing/malnutrition_sp/ (translated by me)

5: What Do You Say to My Plan, "Save All Now and Forever"?

The idea is to intensively invest abundant money NOW into SDGs and the policies of the UNFPA (United Nations Population Fund). The objective of this plan is to eliminate hindrances to SDGs and the UNFPA. At the same time, it resolves the COVID-19 pandemic by making huge investments immediately.

SDGs, as you may already know, are 17 goals set by the United Nations to be met by 2030 that intend to keep the world sustainable. The goals include eliminating hunger and poverty.

Although we are suffering from the COVID-19 pandemic, another present reality is that nearly 14,000 people are dying of hunger daily.

Why shouldn't we, the developed countries, increase our efforts to end the suffering of other people worldwide, including the 800 million suffering from hunger?

Why not, especially if we, the US, EU, and Japan, are capable of protecting our people from the COVID-19 pandemic by investing no less than $2 trillion, €750 billion, and $1 trillion, respectively?

Let's work together to save everyone, collecting money from around the world.

Here is our message:

"Your country first? Okay, but let's save the world altogether.

We can help ourselves and our neighbors at the same time."

Above all, I consider it essential to lead the world to a sustainable and adequate population, where it strikes a balance between demand and supply, harmonized with the environment, by a particular way of investment.

It is to implement an estimated budget by UNFPA of $264 billion.

Isn't this how we can witness the real advent of the peaceful world everyone wants, eliminating hunger and poverty, and achieving an adequate world population?

The investment mentioned above must be quite different from watering a desert in vain.

On the contrary, just by implementing the plan, I believe we can achieve a peaceful world that is like a green field with flowers blossoms beginning to bloom.

6: What Is the Main Cause of Poverty?

Conclusion: We should keep in mind that the liberal economy itself is the main cause of poverty. It should be improved through Basic Income and other means.

Because, In the liberal economy, there is competition, both among firms and among workers, and the weak and the losers in the competition tend to have lower or lost incomes, i.e., they tend to fall into poverty.

Examples of the result of such competition include unsold products, stores that go out of business, bankruptcy, job layoffs, unemployment, people failing better jobs, and people falling into low-wage labor.

I think that you can easily imagine the sufferings, because those sufferings of the unemployed during the COVID-19 disaster are fresh in our minds.

However, this does not mean that I should completely reject a liberal economy and the competition that goes along with it.

Because it is through competition that better goods can be offered at lower prices, workers can compete for better services, and people can enjoy these services.

In a developed country such as Japan, you can find superior goods (because of competition) at reasonable prices, service workers who are well attentive. From foods to appliances, housing and utilities, and so on., people can enjoy such better goods and services. This can easily be understood by simply going to the neighborhood store or looking at what you wear or eat. This is a good aspect of the liberal economy.

Nevertheless, as mentioned earlier, competition brings with winners and losers, with the lower ranks and losers resulting in lower incomes, losses, and bankruptcies for businesses, and lower incomes and unemployment for workers.

Thus, in today's liberal economies of major countries, a situation has arisen in which some people are poor while others are rich, although goods and services are affluent in towns.

One of the measures to improve this situation is a Basic Income.

One way to remedy this situation is to introduce a Basic Income (a Basic Income is a policy in which the government provides a certain amount of benefits to each citizen), because it will ensure the basic livelihood of the less competitive while taking advantage of the benefits of competition without denying it, thereby improving the poverty problem.

Although Basic Income has some unresolved issues such as securing financial resources and international cooperation, we should not give up on introducing it just because it has problems.

Instead of thinking of reasons why it cannot be done, we should consider how it can be done and how the problems and issues can be resolved.

In conclusion, I have argued that a liberal economy creates both affluence and poverty, and that a Basic Income system should be introduced as a safety net in place of the current public assistance.

7: Letter to Avengers

If you are avengers who harm people who you consider are evil to attain your justice,

I must say you are not justified any more.

Please transform yourselves into "educators of love" and love others, even those you consider evil.

For the ultimate peace: Eternal Happiness.

Reasons and how you should do are as follows,

It is clear that to avenge is against the words of Jesus "You must not kill" "love your enemy".

I agree some revenge and punishment as an example are effective to maintain order,

and that hating the evil and seeking justice have brought order to society.

But I oppose to inflict, sadden and even kill any people even in case you consider they are evil.

However, I think that people would joyfully inflict, sadden and kill others who they consider evil, in the name of their justice.

Maybe such people are too influenced by heroic images. We must not do such things.

The creation itself has responsibility for generating the evil.

We are not born good or evil.

It is on the whim of the creator what a being would become after its creation. That must be like random number generation.

Then, what is the right path to follow?

The right path is, to change the world into a world where those lost to evil can be saved and brought back into the light of love!

The penalty or punishment should only be what is needed to maintain order, not so harsh it destroys the individual.

If there exists true justice, it must be "Eternal Happiness" of all people including ones who lost to evil.

It is against our Eternal Happiness to make any person profoundly unhappy.

Not harming yourselves and others are required for our Eternal Happiness to rule our lives.

Great love, even to love your enemy must be necessary this to occur.

Therefore, if you are avengers, please transform yourselves into "educators of love"!

Please exert your mighty force to protect and enlighten every single person.

If a person or a group claims its superior to others, he/they should be superior in such a love as well.

Although I might be very weak to love others, I believe you are the ones who can do it.

For the ultimate peace: Eternal Happiness!

8: Coexistence of Libertarianism and Everyone's Happiness

~Lovers of liberty, let's empower love to take care of the weak~
(This is an open letter)

Dear Libertarians

How about redefining one of your ideals; "the minimal state" as being "minimal enough only to secure people's happiness and freedom?"

Then why not empower both the government minimally and charity organizations to take care of the needy?

In this way, the basis of happiness for all should be secured, even if you would not personally interfere with others and pursue individual liberty.

Firstly because in helping the needy, both the government and non-government charity organizations have their advantages and disadvantages and they should supplement each other.

The advantage of the private sector is that it is "flexible and able to be deeply involved," while its disadvantage is that it is "unstable and weak," compared with the government.

The advantage of government welfare is that it is "stable and powerful" and its disadvantage is that it is "inflexible".

Secondly because without the above-mentioned mutual supplement, the needy won't be adequately provided for and everyone can't be happy.

At first glance, the government helping everyone seems to contradict the "minimal state." Still, by considering that minimal power should be given to the government only for "everyone's freedom and happiness," happiness for all and libertarianism can be compatible.

Lovers of liberty, what do you think of my opinion?

If you agree with me, let's empower love to take care of the weak!

Thank you for reading the entire letter.

I look forward to hearing from you.

Sincerely,

Takeo Omae (My real name)

(IV) About Me

You might think that I have an excellent sense of ethics, since I have written much about peace and happiness, etc. Unfortunately, I do not have such a personality. I do not have an abiding love for my neighbors. I am sorry for that.

I am also sorry for my imperfections and my past mistakes, including unintended consequences. I will try to keep on changing my way for the better.

/ 109

About the Author

I am a Japanese freelance peace activist, born in 1972. My real name is Takeo Omae. I have been interested in peace since childhood. When I was a student at Hosei University, I visited the former Yugoslavia for about two weeks to learn about the conflict. Presently, I run a website about peace, "JC STATION".

Read more at https://www.jcstation.info/e.home.htm.